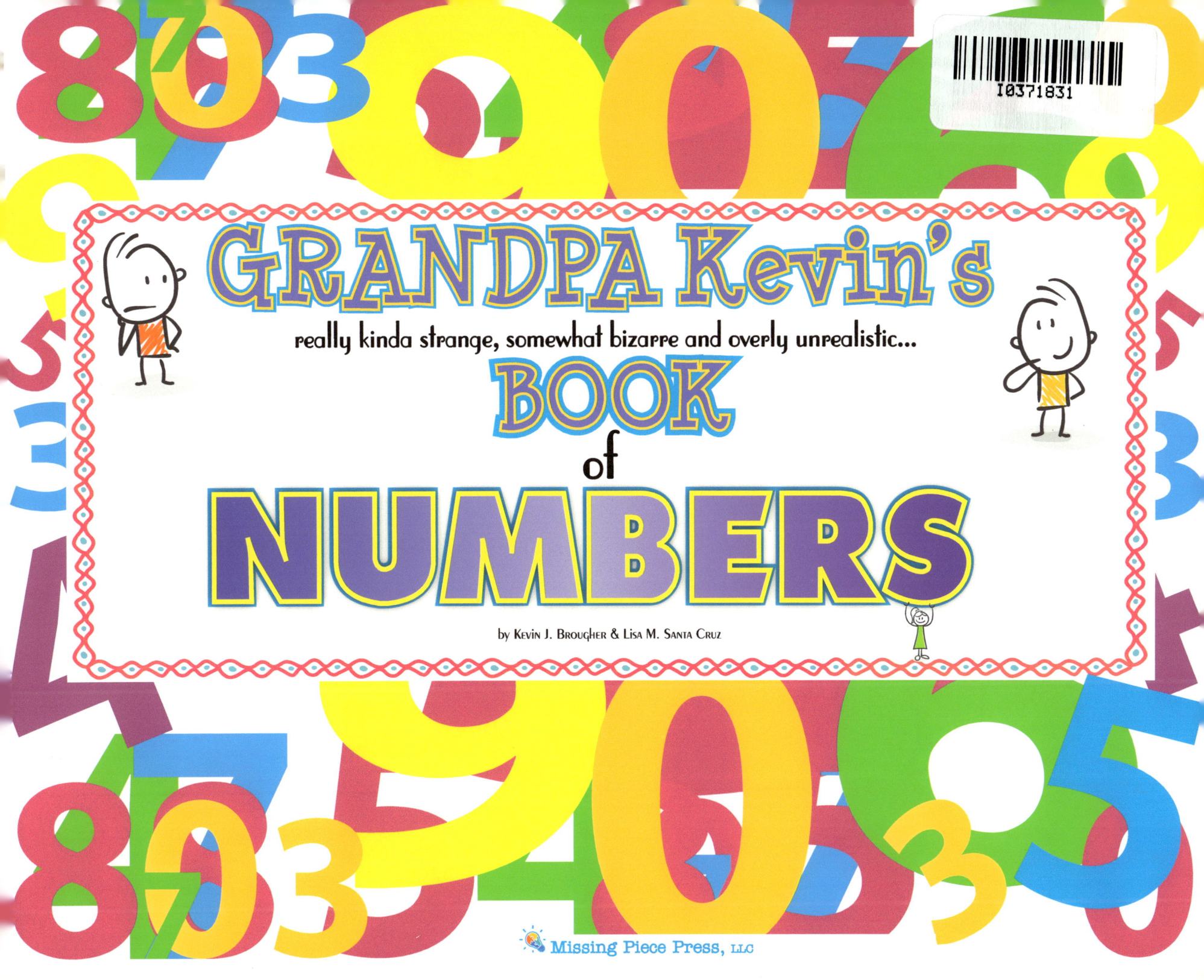

Grandpa Kevin'...Book of Numbers

Copyright © 2020 Kevin J. Brougher & Lisa Santa Cruz

Hardback ISBN# 978-0-9977959-9-8
Paperback ISBN# 978-1-957035-07-9
Library of Congress Control Number : 2019913812

All rights reserved.
No part of this book may be reproduced by any means
without the written permission of the publisher.
Printed in U.S.A.

Distributed in United States and Canada by Allied Resources - Abilene Texas.
For wholesale and licensing inquiries, please email
us at : Questions@MissingPiecePress.com

Missing Piece Press, LLC does not accept unsolicited manuscripts.
: MissingPiecePress.com for information on our other fun products.
LIKE us on Facebook to keep up to date with our new books, games and special offers.
"A little Thinking...a LOT of FUN!" is a trademark of Missing Piece Press, LLC.

*Missing Piece Press is a publisher
of award-winning books and games.
Our goal is to produce products that fill the user
with a sense of fun, wonder, and intrigue.*

A Little Thinking...a LOT of FUN!®

Other Publications from Missing Piece Press

BOOKS

Thinklers! 1 : *A Collection of Brain Ticklers!*
Thinklers! 2 : *More Brain Ticklers!*
Thinklers! 3 : *Even More Brain Ticklers!*
Thinklers! 4 : *Full-Color Brain Ticklers!*
History Mysteries : *A New Twist on Time-Lines*
State Debate : *50 Unique Playing Cards and 50 Games*
Number Wonders : *A Collection of Amazing Number Facts!*
Dreams, Screams, & JellyBeans! : *Poems for All Ages*
The Storybook : *A novel for ages 10 on up*
Science Stumpers : *Brain-Busting Scenarios...Solved with Science*
Algebra Summary Sheets : *Posters to Promote Proficiency*
Reindolphins : *A Christmas Tale*
Who Says Hoo? : *A Book for Babies & Toddlers*
Grandpa Kevin's... ABC Book : *really kinda strange...*
Grandpa Kevin's... Book of COLOR : *really kinda strange...*
Number Fun! : *A Book of Counting and Numbers for Toddlers*
Who's Waiting for You? : *A Book of Animal Clues for Toddlers*
Night Owl : *A Book for Nocturnal Toddlers*

GAMES

Frazzle : *A Frenzied Game of Words*
ShanJari : *An African Game of Sequence and Strategy*
Whew! : *Words, Wits, Whims & Woes!*
TooT! : *A Nerdy Little Game*
Blam! : *A Different Card Game*
DICE Blam! : *A Different Dice Game*
Word Nerd : *A Quick-Witted Word Game*
Bunco BUDDIES! : *The BETTER Bunco Game*

and MORE!

Missing Piece Press, LLC
Copyright © 2020 Missing Piece Press, LLC MissingPiecePress.com

Introduction

For all the families
 who know in their minds
that **ALL** young kids
 will eventually find
that numbers explain
 SO many things -
from huge black holes -
 to quantum strings.
So, if they're ready
 to think and engage -
hurry up!   Turn the page!

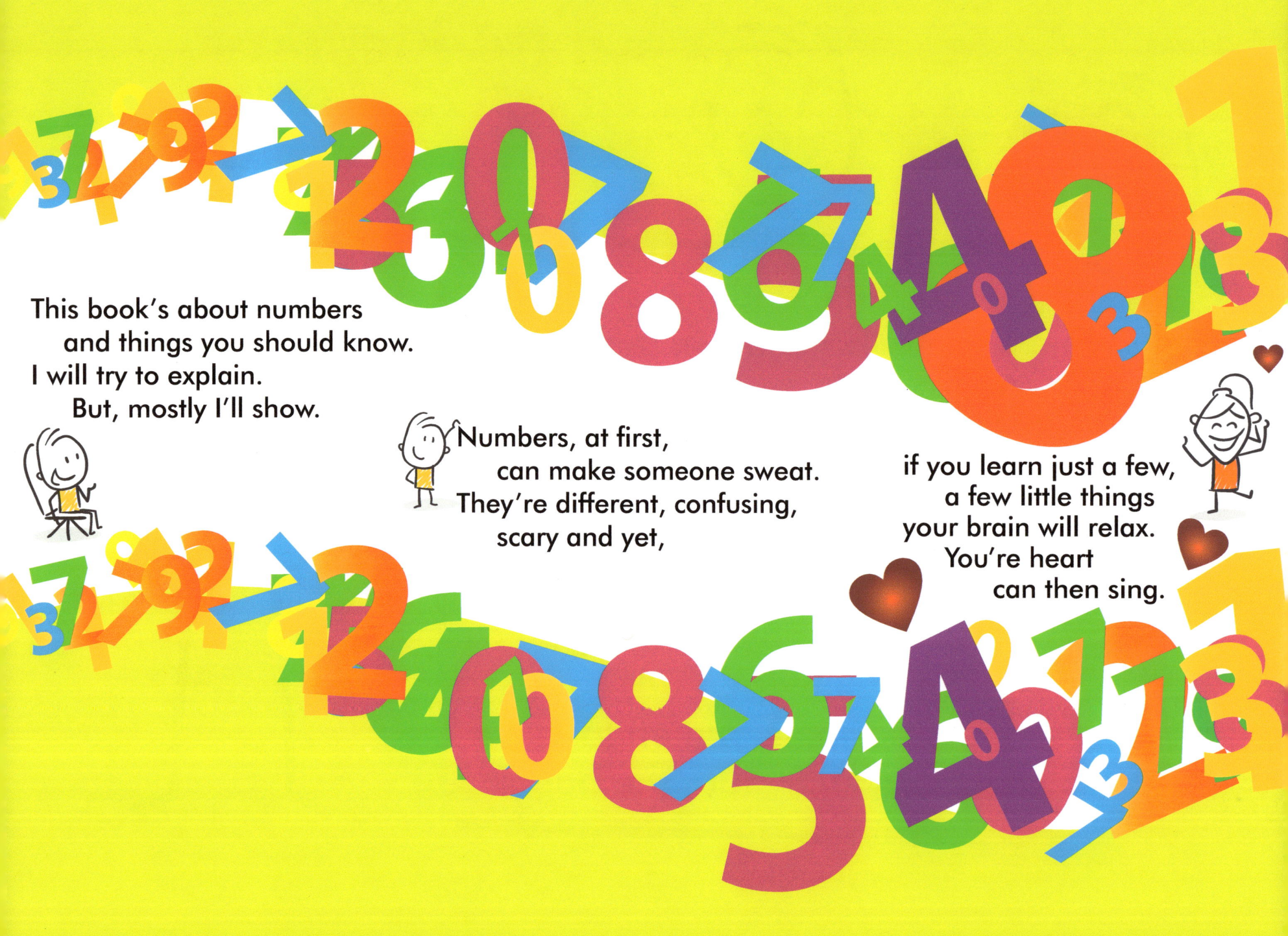

This book's about numbers
and things you should know.
I will try to explain.
   But, mostly I'll show.

Numbers, at first,
   can make someone sweat.
They're different, confusing,
   scary and yet,

if you learn just a few,
   a few little things
your brain will relax.
You're heart
   can then sing.

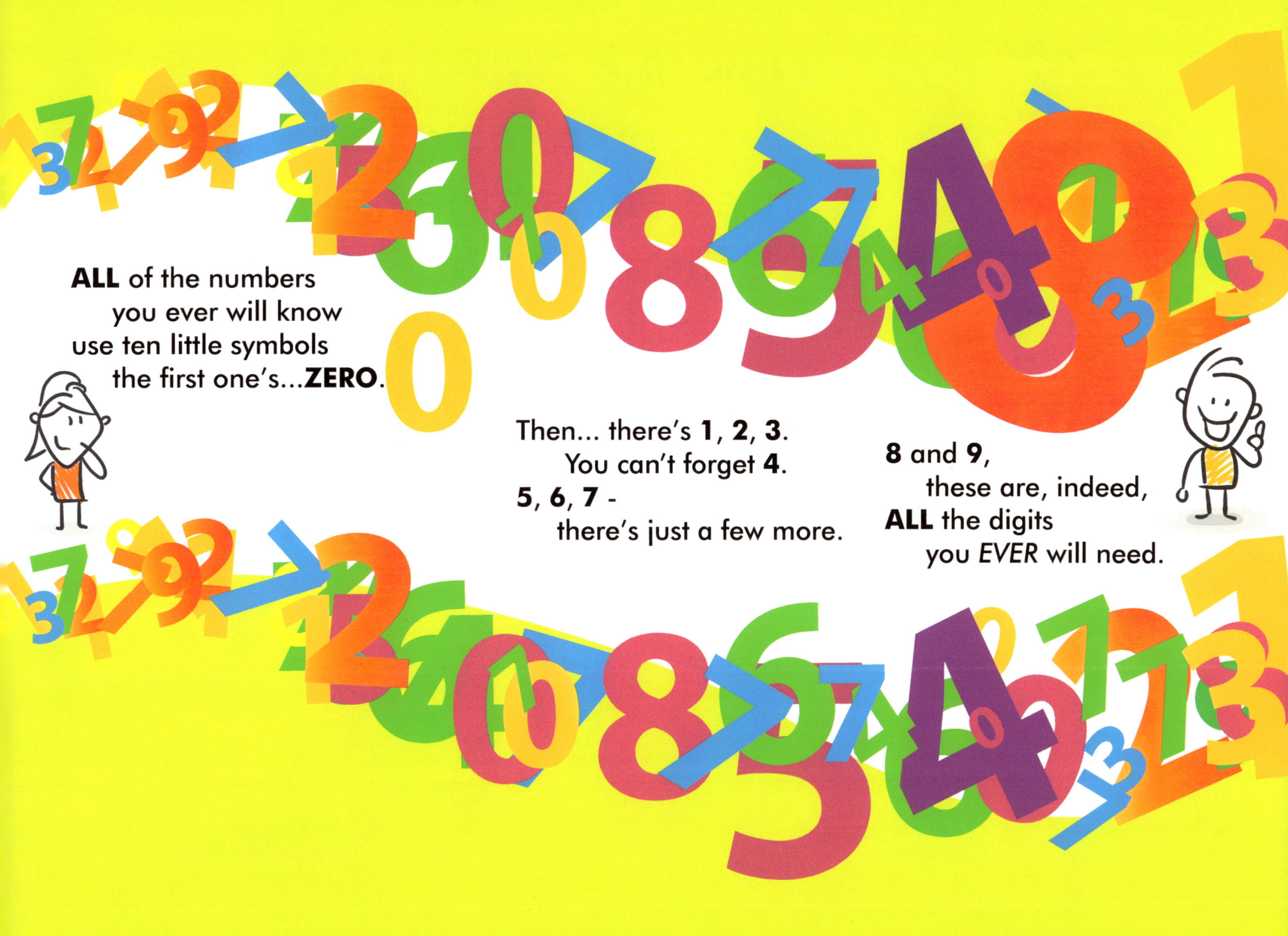

ALL of the numbers you ever will know use ten little symbols the first one's...**ZERO**.

Then... there's **1, 2, 3**.
You can't forget **4**.
5, 6, 7 -
  there's just a few more.

8 and **9**,
  these are, indeed,
ALL the digits
  you *EVER* will need.

Now, the AMAZING thing,
if you're able to read
three digits together
then you can succeed
at reading six digits,
or SO many more.
It soon will be easy -
no longer a chore.

But, here is the list of the **HARDEST**, I think - the **HARDEST** numbers. They really do stink.

*They STINK because you don't read the TENS place and then the ONES place, like we usually do. You have to combine them into these single words.

11	12	13	14	15	16	17	18	19
eleven	twelve	thirteen	fourteen	fifteen	sixteen	seventeen	eighteen	nineteen

But, once you know them you know them for life. No more stress. No more strife.

When we see, five - zero.
We read it as **FIFTY.**
We **SHOULD** say five **TENS**.
Now, that would be nifty!

But, wouldn't you know,
call it bad luck,
they made up a rule
and somehow it stuck.

So, before we go on you must be an ace at reading the digits when in the **TENS** place.

?	**?**
TENS	**ONES**

20	30	40	50	60	70	80	90
twenty	thirty	forty	fifty	sixty	seventy	eighty	ninety

So, read the **TENS,** then read the **ONES.**
So after **TWENTY** (20), is **TWENTY** - **ONE** (21).
There's 22 **(twenty two),** then 23 **(twenty three).**
It's not too hard - I guarantee.

TRY iT!

27 - **TWENTY** seven
33 - **THIRTY** three
48 - **FORTY** eight
51 - **FIFTY** one
67 - **SIXTY** seven
79 - **SEVENTY** nine
80 - **EIGHTY**
94 - **NINETY** four

Now, we're ready
  for the three digit game.
The three digits you read
  are like a **FIRST** name.

There's **HUNDREDS** and **TENS**
  and, of course, there are **ONES**.
Now, let me show you
  just how it's done.

Read left to right - like reading a book. The HUNDREDS place is the first place to look.

❓ ❓ ❓
Hundreds Tens Ones

This place will always tell you how many HUNDREDS you have. A HUNDRED is **10**, TENS.

This place will tell you how many TENS you have. The TENS have **DIFFERENT** names (as we discussed). We ***SHOULD*** say 3 ten, 5 ten, or 9 ten just like we say 3 hundred, 5 hundred or 9 hundred. But, instead, we say **THIRTY** (30), **FIFTY** (50), or **NINETY** (90) etc.. Don't ask me why.

* And, as we discussed, if there is a 1 in the TENS place and any number, other than zero, it will be a *"TEEN"* number. Which is stinky! Really stinky!

This last place will always tells us how many SINGLES or ONES we have.

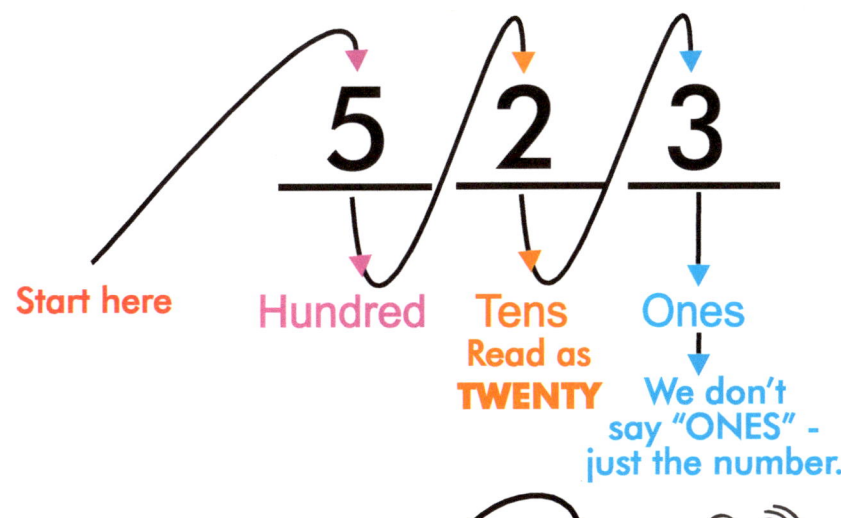

So, this number is:
Five HUNDRED TWENTY three

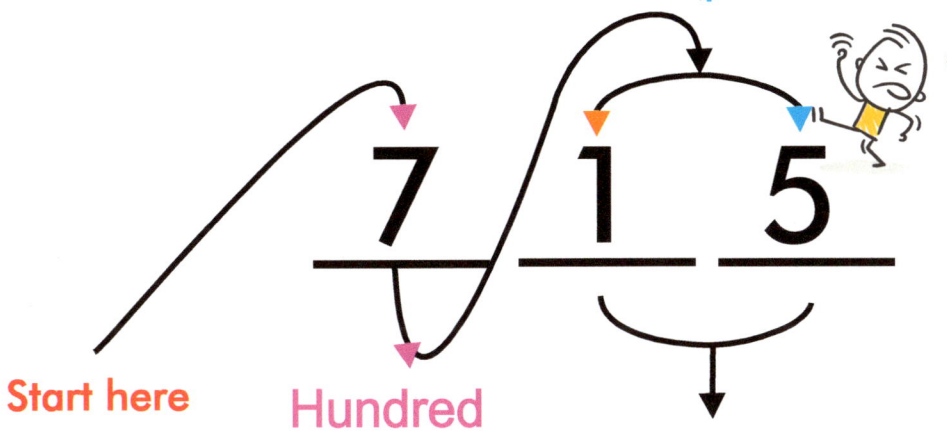

STINKY number!

So, this number is:
Seven HUNDRED Fifteen

Uh...oh! Remember the HARDEST numbers I mentioned?! THIS is one of them! You read the TEN and ONES *TOGETHER* as **FIFTEEN**.

TRY IT!

Try reading the number on your own and THEN checking to see if it matches what is written.

381 — Three **HUNDRED** - Eighty (Eight **TEN**), **ONE**

245 — Two **HUNDRED** - Forty (Four **TEN**), **FIVE**

902 — Nine **HUNDRED** - (No **TENS!**) - **TWO**   *NOT Nine hundred **AND** two

114 — One **HUNDRED** - **FOURTEEN**   *Tricky!*

570 — Five **HUNDRED** - **SEVENTY** - (No ones!)

A common mistake, one that is wrong is to say the word, "**AND**" where it doesn't belong.

The **ONLY** time to take *THAT* action is **IF** the number has a fraction.

902.3
Nine hundred two **AND** three tenths

902 1/3
Nine hundred two **AND** one third

But, this book is really **NOT** about fractions and decimals so, let's get back to the *"regular"* numbers.

Reading three digits
   is a wonderful feat.
Especially since
   the process **repeats!**

You just have to add
   all the "**Last Names**"
to complete this challenging
   number fun game!

But, wouldn't you know
   one group breaks the rules.
This one set of three...
   should go back to school!

ALL of the groups have a last name, except, the set on the right... they're somewhat inept.

But, no worries, no worries just read it as normal. It won't have a last name - it won't be as formal.

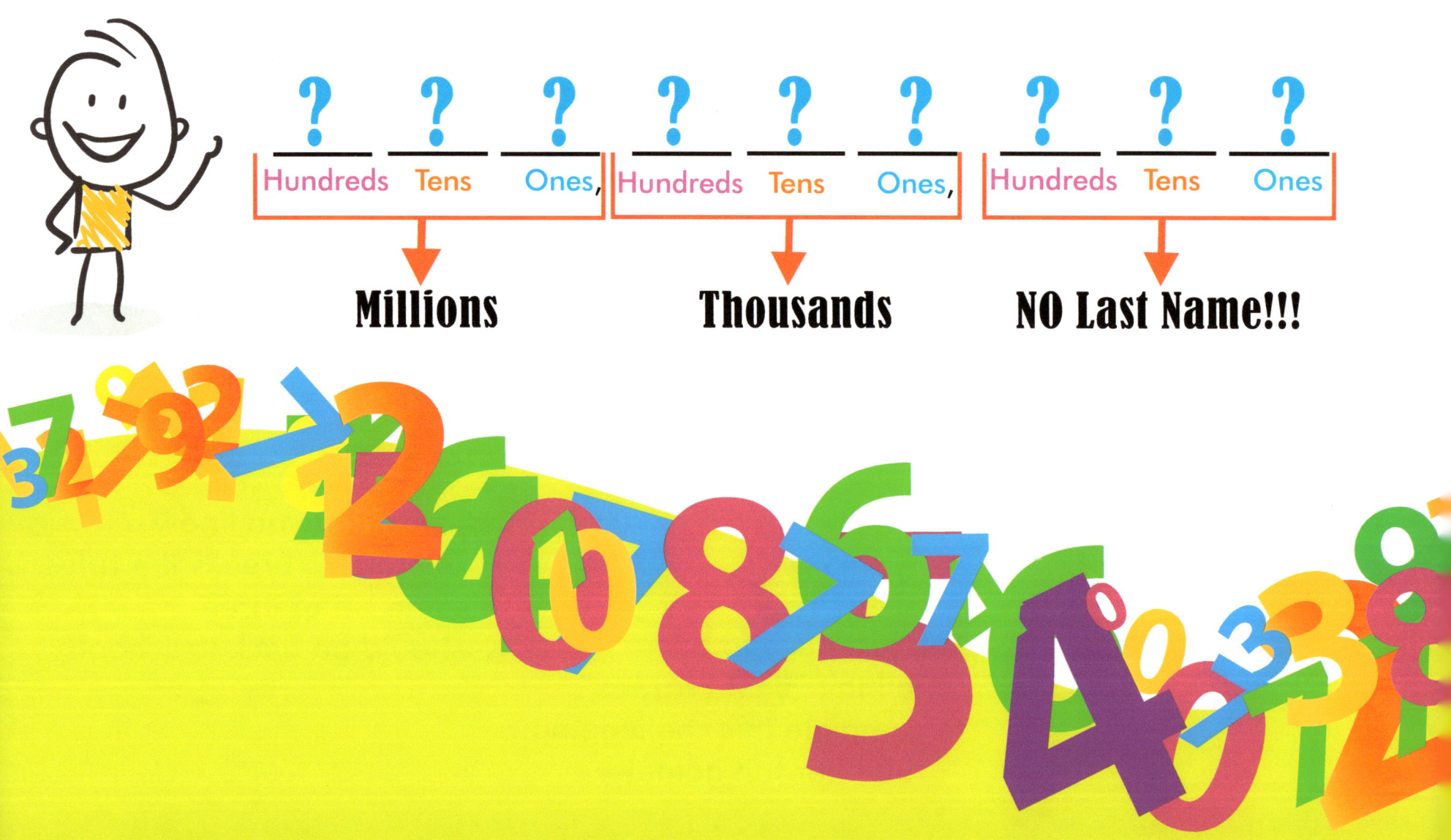

| Hundreds | Tens | Ones, | Hundreds | Tens | Ones, | Hundreds | Tens | Ones |

Millions   **Thousands**   **NO Last Name!!!**

So, read this number,
   just like we've done.
The number here is -
   three hundred one.

301,739

THOUSAND

Now after that
   the word you say
is **THOUSAND** and
   that's **MUST** - not MAY.

Read the last
   3 digits, and then,
you've read it all -
   you've reached the end!

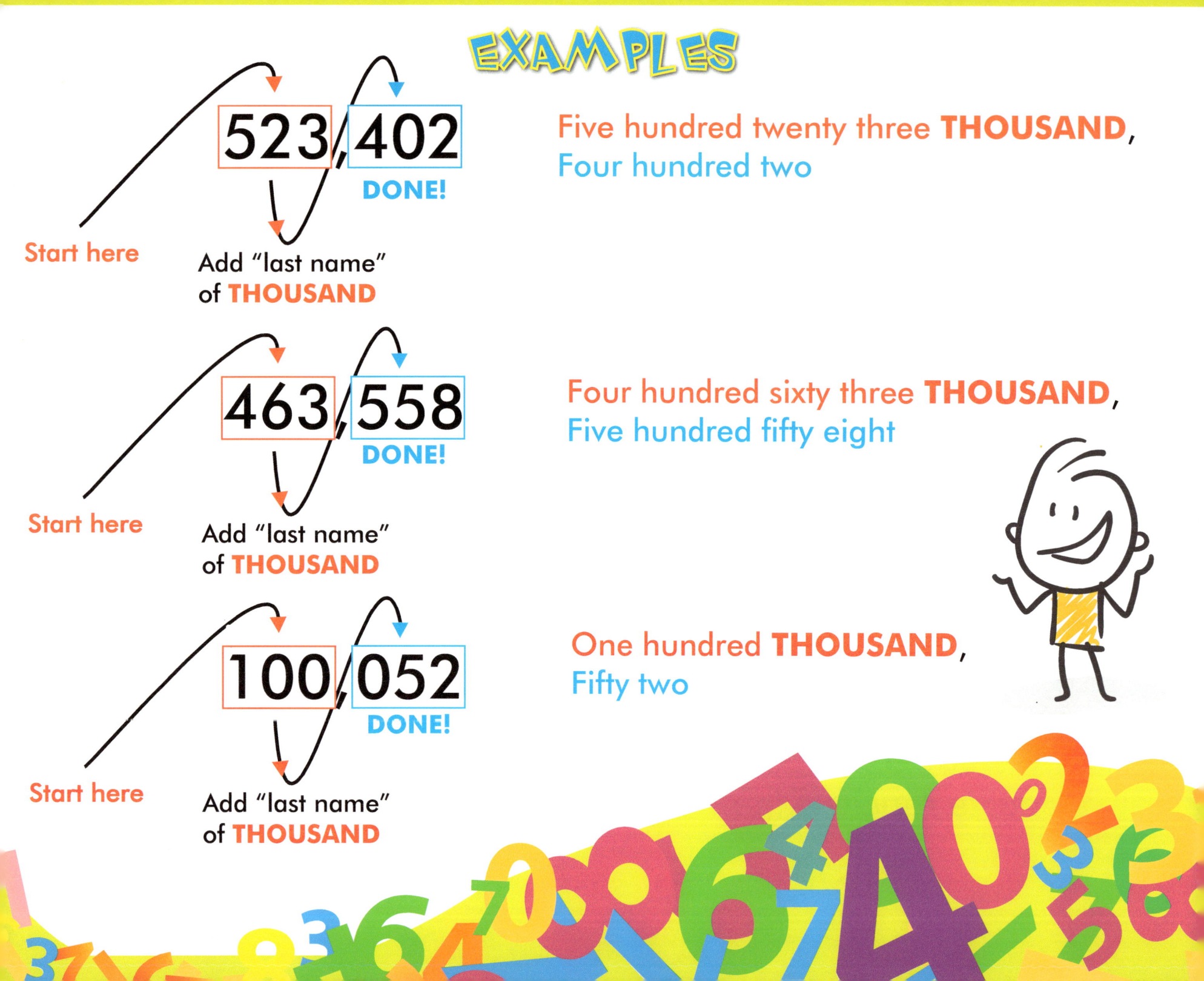

Try reading the number on your own and THEN checking to see if it matches what is written.

541,893 Five hundred forty one **THOUSAND**, Eight hundred ninety three

123,456 One hundred twenty three **THOUSAND**, Four hundred fifty six

709,024 Seven hundred nine **THOUSAND**, Twenty four

Now you're ready
   for a third set of three.
It's not much harder.
   I'm sure you will see.

MILLIONS is
   the term we will use.
You can't ignore it.
   You can't refuse.

So, as you've done
    just read the three -
add the last name (**MILLIONS**)
    and you will see
that when you read
    the **THOUSANDS** and
the last three digits
    at the end,
you will have read
    a big, big number.
You'll be proud
    and filled with wonder!

EXAMPLES

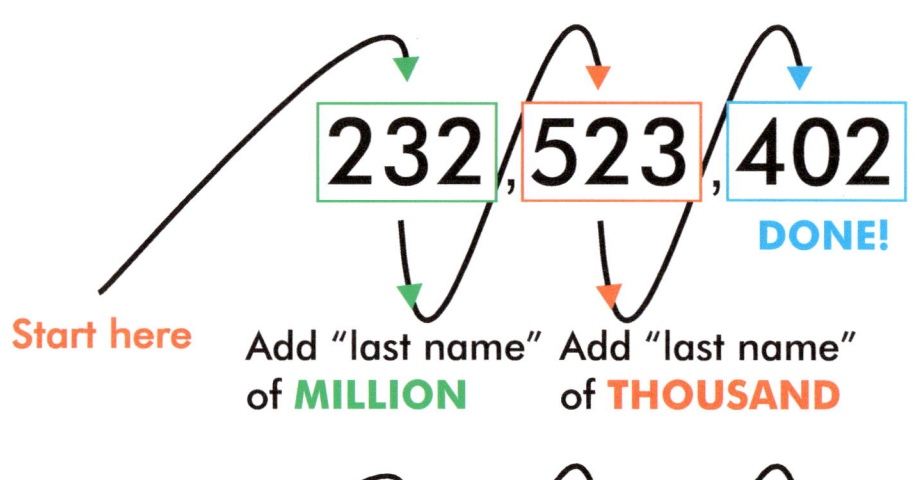

Two hundred thirty two **MILLION**
Five hundred twenty three **THOUSAND**
Four hundred two

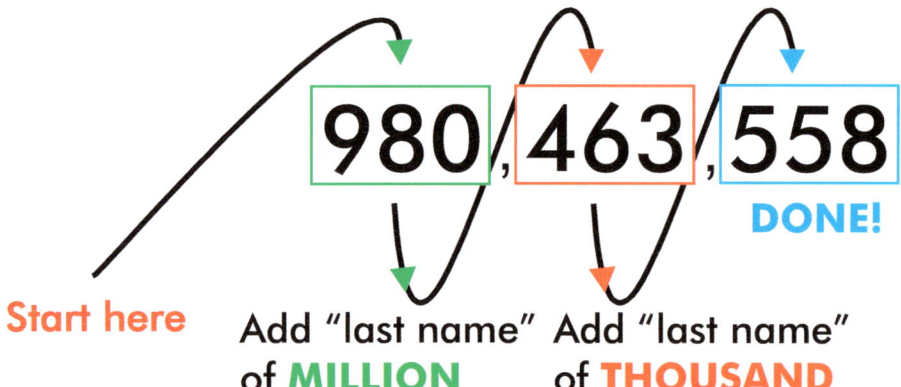

Nine hundred eighty **MILLION**
Four hundred sixty three **THOUSAND**
Five hundred fifty eight

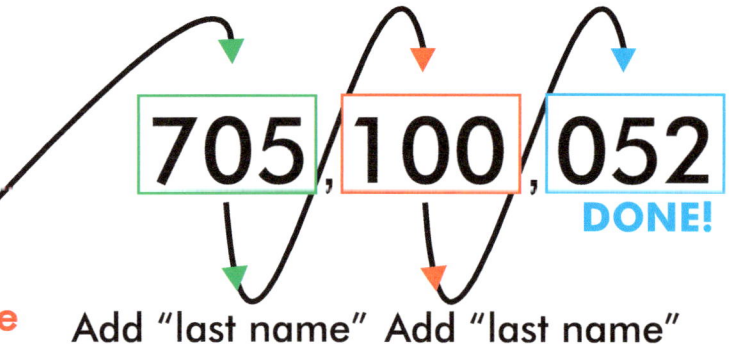

Seven hundred five **MILLION**
One hundred **THOUSAND**
Fifty two

But, you can't deny. You can't dismiss that we must, we must PRACTICE this!

TRY iT!

Try reading the number on your own and THEN checking to see if it matches what is written.

939,528,316

Nine hundred thirty nine **MILLION**
Five hundred twenty eight **THOUSAND**
Three hundred sixteen

78,801,021

Seventy eight **MILLION**
Eight hundred one **THOUSAND**
Twenty one

702,003,007

Seven hundred two **MILLION**
Three**THOUSAND**
Seven

MILLIONS and **THOUSANDS**
 and the three at the end.
I think you are ready.
 It's time to extend
your knowledge of numbers
 so, the next term to know
is **BILLIONS**, yes, **BILLIONS**.
 Ready? Let's go!

___ ___ ___, ___ ___ ___, ___ ___ ___, ___ ___ ___
    BILLIONS   **MILLIONS**   **THOUSANDS**
        ↑

It's the 4th set of three!

EXAMPLES

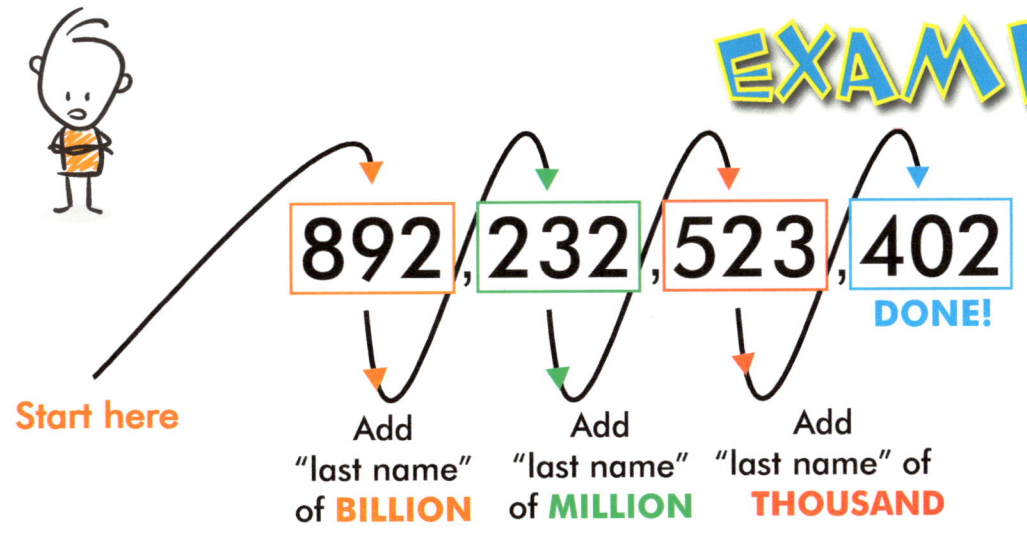

Eight hundred ninety two **BILLION**
Two hundred thirty two **MILLION**
Five hundred twenty three **THOUSAND**
Four hundred two

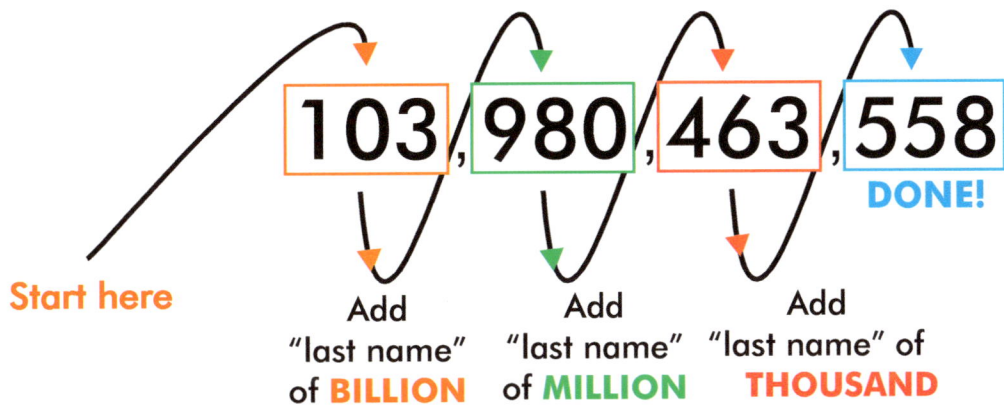

One hundred three **BILLION**
Nine hundred eighty **MILLION**
Four hundred sixty three **THOUSAND**
Five hundred fifty eight

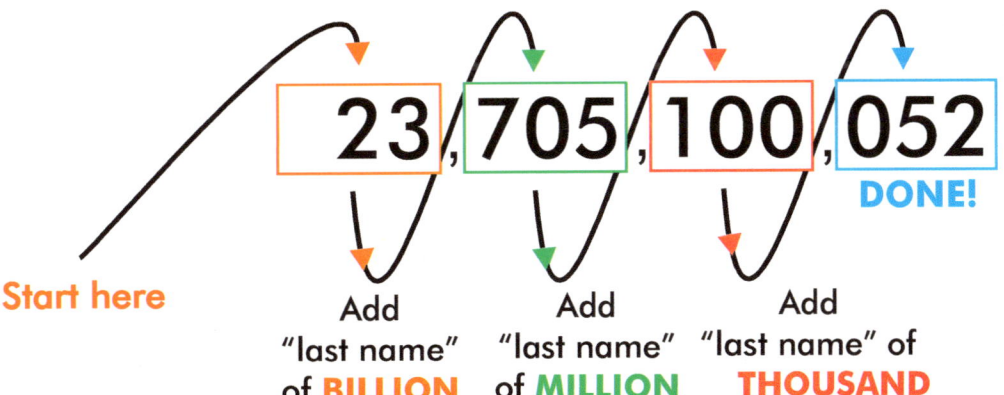

Twenty three **BILLION**
Seven hundred five **MILLION**
One hundred **THOUSAND**
Fifty two

But, you can't deny. You can't dismiss that we must, we must PRACTICE this!

Try reading the number on your own and THEN checking to see if it matches what is written.

12,939,528,316

Twelve **BILLION**
Nine hundred thirty nine **MILLION**
Five hundred twenty eight **THOUSAND**
Three hundred sixteen

455,078,801,021

Four hundred fifty five **BILLION**
Seventy eight **MILLION**
Eight hundred one **THOUSAND**
Twenty one

108,702,003,007

One hundred eight **BILLION**
Seven hundred two **MILLION**
Three **THOUSAND**
Seven

There's just one more last name
   or term we will learn.
But, I have small thought.
   I have a **CONCERN**.

There often are numbers
   where zero is queen.
When you look at the number
   in most places - it's seen.

So, when zero's the hero
and overly ample,
here's what to do -
here's an example :

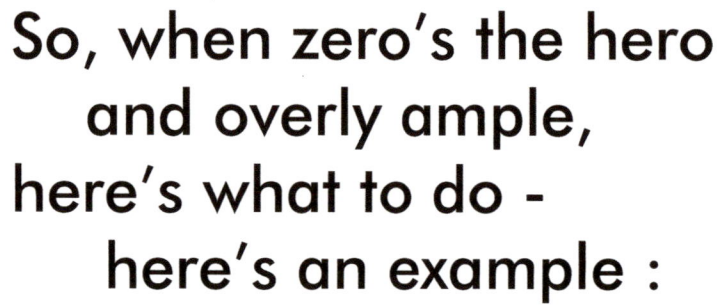

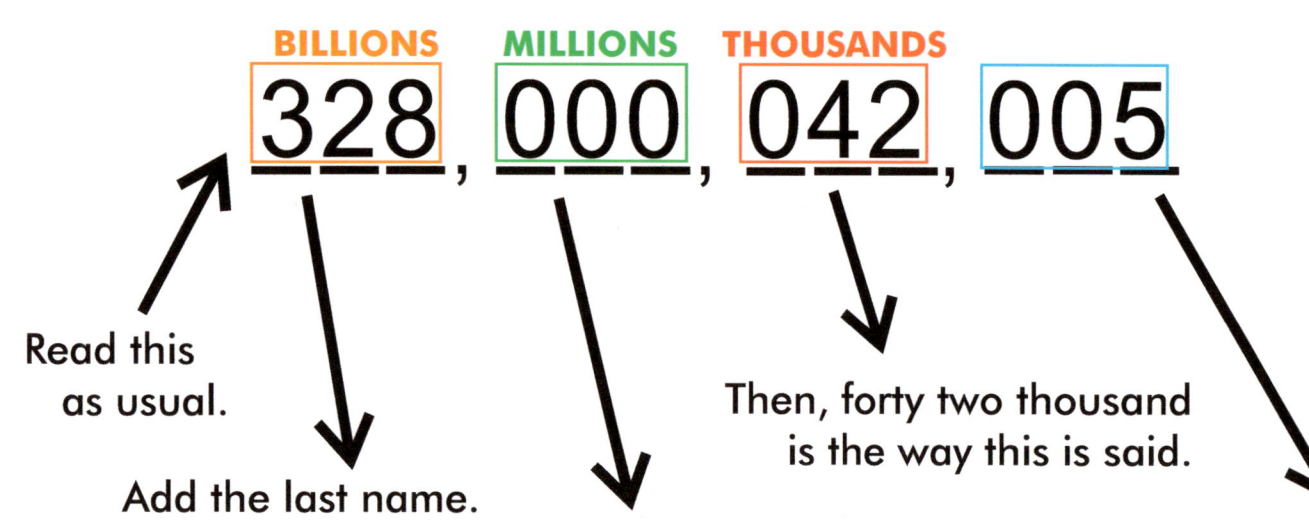

BILLIONS **MILLIONS** **THOUSANDS**

328 , 000 , 042 , 005

Read this as usual.

Add the last name.

Nothing said here,
That's part of the game.

Then, forty two thousand
is the way this is said.

Then, five - just five
is the way THIS is read.

Three hundred twenty eight **BILLION**,
Forty two **THOUSAND**,
Five

And this one :

2,002,000,000
BILLION   **MILLION**   **THOUSAND**

That's just : Two **BILLION**, Two **MILLION**

Or, this one:

5,000,000,004
BILLION   **MILLION**   **THOUSAND**

That's just : Five **BILLION**, Four

TRY iT!

52,000,008,006

Fifty two **BILLION**
Eight **THOUSAND**
Six

400,100,000,021

Four hundred **BILLION**
One hundred **MILLION**
Twenty one

1,000,000,100

One **BILLION**
One hundred

Now, for the last -
   the last name of the day.
Here's what it is -
   here's what to say.

When you get to a number
   with this many digits
It can make people shiver.
   It can make people fidget.

But, **TRILLION**'s not hard.
   It's just like the rest.
Reading a **TRILLION**,
   will surely impress!

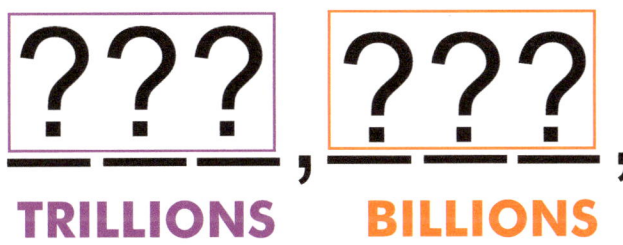

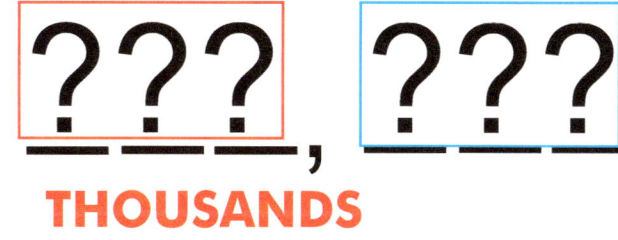

???,???,???,???,???

TRILLIONS   **BILLIONS**   **MILLIONS**   **THOUSANDS**

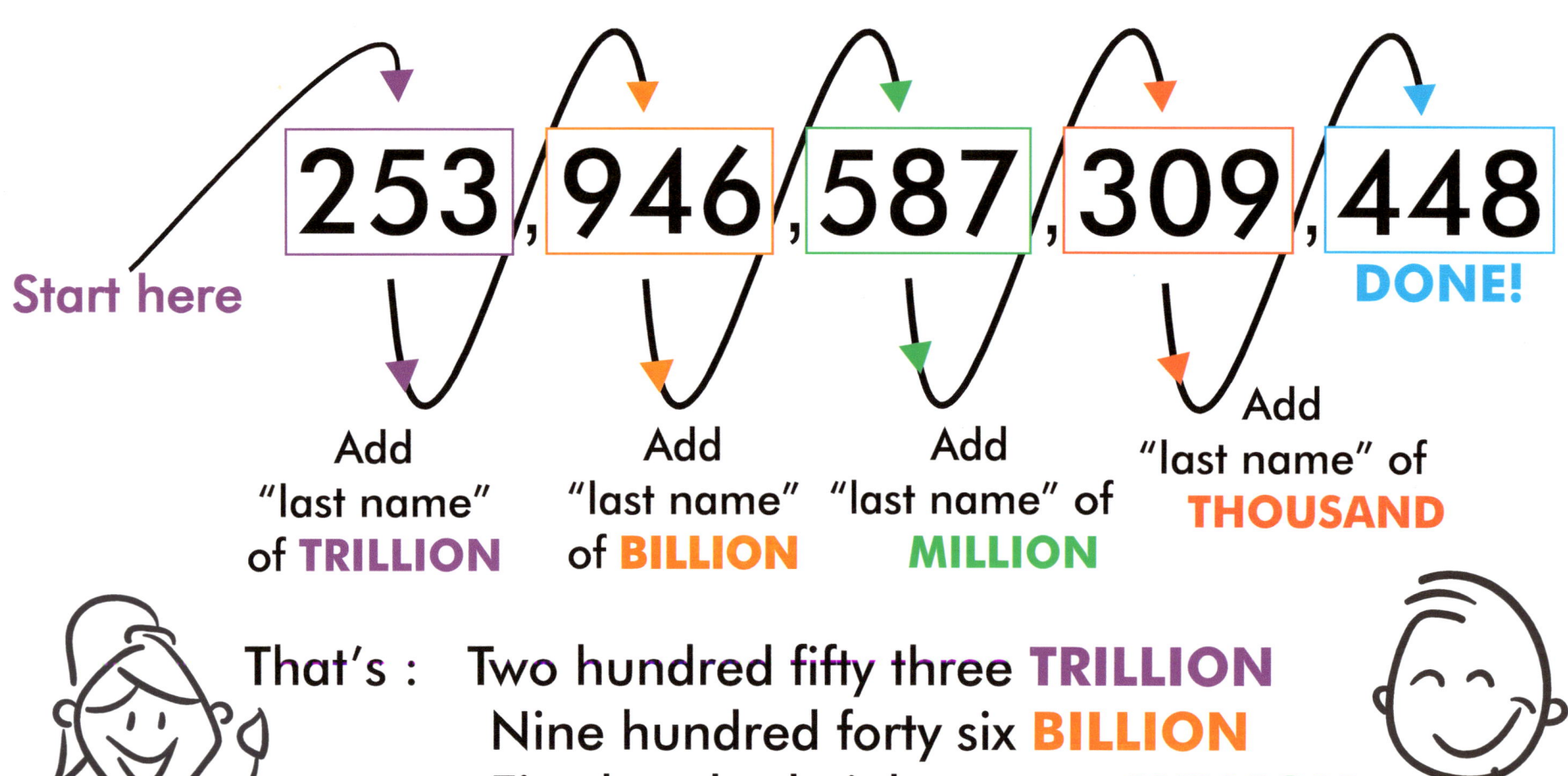

Now with no arrows
  but still a few labels
see if you can -
  see if you're able
to read these big numbers
  all the way through.
If you get stuck,
  take time to review.

Try reading the number on your own and THEN checking to see if it matches what is written.

TRILLIONS BILLIONS MILLIONS THOUSANDS
106,512,939,528,316

One hundred six **TRILLION**
Five hundred twelve **BILLION**
Nine hundred thirty nine **MILLION**
Five hundred twenty eight **THOUSAND**
Three hundred sixteen

TRILLIONS BILLIONS MILLIONS THOUSANDS
987,455,078,801,021

Nine hundred eighty seven **TRILLION**
Four hundred fifty five **BILLION**
Seventy eight **MILLION**
Eight hundred one **THOUSAND**
Twenty one

TRILLIONS BILLIONS MILLIONS THOUSANDS
212,108,702,003,007

Two hundred twelve **TRILLION**
One hundred eight **BILLION**
Seven hundred two **MILLION**
Three **THOUSAND**
Seven

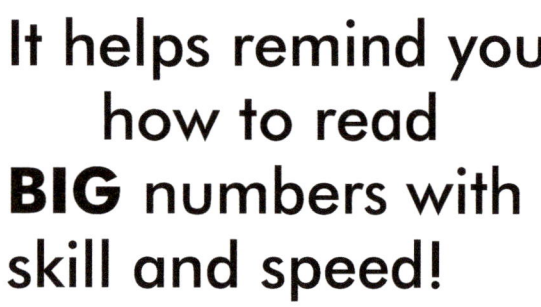

Though you have come
to this book's end,
you should read it
now and then.

It helps remind you
how to read
BIG numbers with
skill and speed!

But, since you made it
this far through,
I have to say...
we're **PROUD** of you!

The END

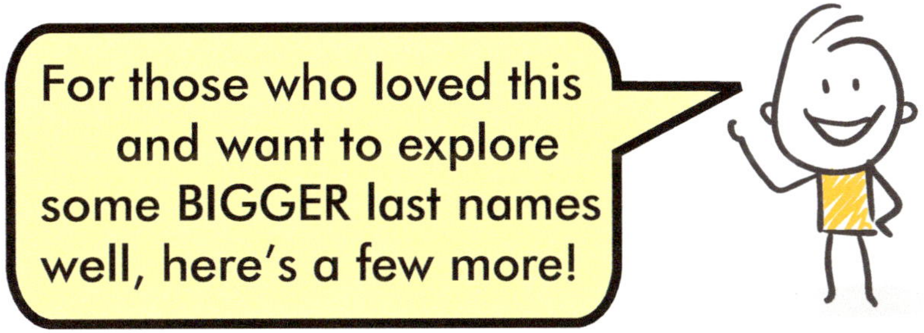

For those who loved this and want to explore some BIGGER last names well, here's a few more!

____,____,____,____,____,____,____,____,____,____,

NONILLION OCTILLION SEPTILLION SEXTILLION QUINTILLION QUADRILLION TRILLION BILLION MILLION THOUSAND

Missing Piece Press, LLC

For our full line of Children's Books, Thinking / Learning Books & Games please visit:
MissingPiecePress.com

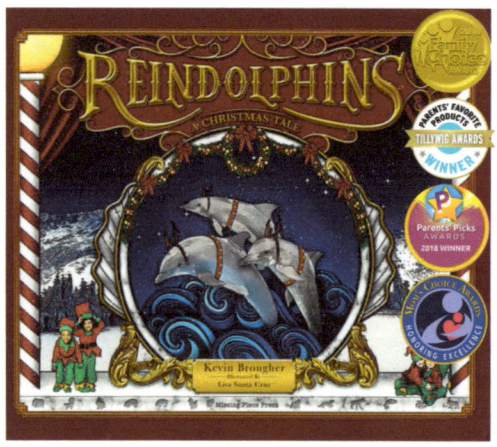
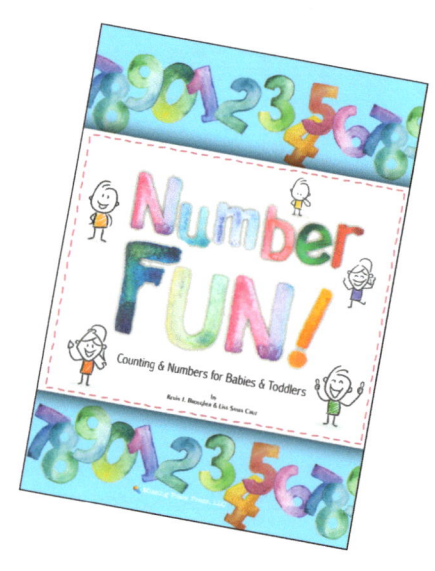

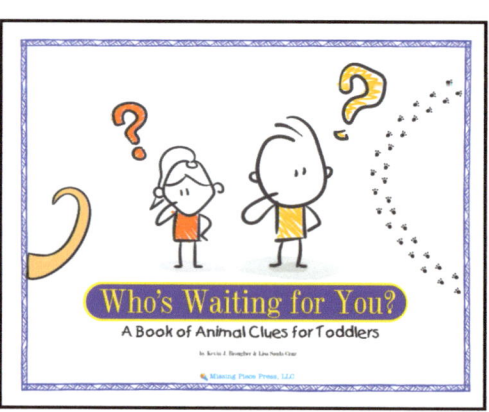

Missing Piece Press, LLC

www.ingramcontent.com/pod-product-compliance
Lightning Source LLC
Chambersburg PA
CBHW041149070526

44579CB00005B/55